*The Most Dammed
Country in the World*

23/05/21

5:12

Engineering Department

1. Greta Thunberg *No One Is Too Small to Make a Difference*

2. Naomi Klein *Hot Money*

3. Timothy Morton *All Art is Ecological*

4. George Monbiot *This Can't Be Happening*

5. Bill McKibben *An Idea Can Go Extinct*

6. Amitav Ghosh *Uncanny and Improbable Events*

7. Tim Flannery *A Warning from the Golden Toad*

8. Terry Tempest Williams *The Clan of One-Breasted Women*

9. Michael Pollan *Food Rules*

10. Robin Wall Kimmerer *The Democracy of Species*

11. Dai Qing *The Most Dammed Country in the World*

12. Wangari Maathai *The World We Once Lived In*

13. Jared Diamond *The Last Tree on Easter Island*

14. Wendell Berry *What I Stand for Is What I Stand On*

15. Edward O. Wilson *Every Species is a Masterpiece*

16. James Lovelock *We Belong to Gaia*

17. Masanobu Fukuoka *The Dragonfly Will Be the Messiah*

18. Arne Naess *There is No Point of No Return*

19. Rachel Carson *Man's War Against Nature*

20. Aldo Leopold *Think Like a Mountain*

The Most Dammed Country in the World

DAI QING

PENGUIN BOOKS — GREEN IDEAS

PENGUIN BOOKS

UK | USA | Canada | Ireland | Australia
India | New Zealand | South Africa

Penguin Books is part of the Penguin Random House group of companies
whose addresses can be found at global.penguinrandomhouse.com.

First published in *The River Dragon Has Come!* by Routledge in 1998 and
by *Probe International* in 2002, 2010 and 2011.
This selection published in Penguin Books 2021

001

'The Three Gorges Project' is reproduced by arrangement with
Taylor & Francis Books UK from *The River Dragon Has Come*,
1st Edition by Dai Qing; John G. Thibodeau; Michael R. Williams;
Yi Ming; Audrey Ronning Topping, published by Routledge.
The other five essays are reproduced by kind permission of
Probe International journal.

Copyright © Dai Qing, 2021

Set in 12.5/15pt Dante MT Std
Typeset by Jouve (UK), Milton Keynes
Printed and bound in Great Britain by Clays Ltd, Elcograf S.p.A.

The authorized representative in the EEA is Penguin Random House Ireland,
Morrison Chambers, 32 Nassau Street, Dublin D02 YH68

A CIP catalogue record for this book is available from the British Library

ISBN: 978–0–241–51459–7

www.greenpenguin.co.uk

Contents

China's 'Rise' and the Environment's
Decline   1

The Three Gorges Project   23

The race to salvage the Three
Gorges treasure trove   46

Who wants this dam anyway?   55

Is 'keeping in step with the Party'
good for the environment?   60

The Yangtze and the Three Gorges   66

China's 'Rise' and the Environment's Decline*

The whole world is talking about 'China's rise.' Even the Chinese people themselves – especially officials and the official media – describe the current situation as the achievement of 'a prosperous society' brought about by 'the miracle of economic growth.'

We behold China's annual GDP growth and the government's stash of foreign bonds.

We behold skyscrapers and the Bird's Nest Stadium.

We behold the largest dam in the world, the Three Gorges project.

*A speech given at the University of Toronto's Munk School of Global Affairs and Public Policy on 26 October 2010.

Behold the country's massive export of toys and electrical appliances.

China is also looking to space and has launched its second unmanned lunar probe.

Not so obvious, however, is that China has another very special 'export': the ideology of authoritarianism – a very special export that feeds China's 'rise' and makes China seem even more powerful.

The most attractive new faces advertising 'China's success today' – the poster children of a 'rising China' – are the new rich Chinese who have emerged in China and elsewhere in the world over the past 20 years. These people are lavish, smart, and arrogant. They feel they can do anything they want, and that there is nothing they cannot do.

In China, they are known as the newborn Red Nobility. The Red Nobles are government officials and their family members, or at least those with very strong links to these government officials (former secretaries, etc.) who have their careers in China.

They are conspicuously wealthy, elite and self-confident.

Recently, netizens* have coined a term for the new special elite called 'Naked Officials.' These 'Naked Officials' move their cash and their dear ones – wife, children, concubines – abroad, buying houses and cars for them in their new countries of residence. Meanwhile, these 'Naked Officials' continue to live in China but usually have several passports in hand and are prepared to escape China at any moment to join their families (and lovers) abroad.

Is this the evidence of China's 'rise'?

We must ask a few questions about this 'miracle' of growth and prosperity:

Is this 'rise' like manna from heaven?

Or does it come at a cost to the Chinese people and the world? If so, what is that price?

---

* Unable to participate openly in civic matters, Chinese citizens interact with each other and express their opinions and hold debates on the Internet, hence their description as 'netizens.'

Before answering this question, we must have a look at today's People's Republic of China. What kind of country is it in the modern age?

What is the nature of its 'rise,' and why didn't the 'rise' occur during the Qing Dynasty in the nineteenth century when the Industrial Revolution brought prosperity to many countries?

Why didn't the 'rise' occur at the end of World War II, when China joined the ranks of the victors as one of the 'Four Allies'?

Why didn't the rise occur in Mao Zedong's era, when 'the People became the masters of their own country'?

And why didn't it occur in the 1980s, when the Communist Party of China began to rethink its many errors, to relax its grip on society and to implement the policy of 'reform and opening up'?

Why has China only 'risen' in the past two decades, after the tanks ran over Tiananmen Square and shocked the people of China into a terrified silence at all levels of society: high-level party officials, scholars and professors, local officials and cadres, state-owned enterprise managers and workers, private business owners, and students currently in school or recently graduated. Every-one was shocked into silence.

If the ruling party could order the People's Army to fire on the people, what else could it do? People asked, what kind of 'People's Republic' are we? At this zero hour, then top leader Deng Xiaoping gave the answer – saying that, yes, 1,000 have been killed at Tiananmen, but, 'I believe it would be worthwhile to kill 200,000 to buy twenty years of stability for the regime.'

The Tiananmen massacre set the stage for the Red Nobles' political reform – only it was not the reform that the students and other demon-strators imagined but the reform that Deng Xiaoping would soon introduce. He announced this reform during his tour of southern China in 1992. In the new commercial city of Shenzhen he dismissed labels such as capitalism and social-ism and declared that 'development is first thing first' and 'let some people get rich first.'

Deng's declaration freed the leadership from the paralysis that had gripped the country since Tiananmen.

The ruling elite suddenly knew that as public

servants they were really free to serve themselves, as long as they laid low and followed the 'lie low rule'– which they describe as 'use your power as soon as possible before it expires.'

But does this political system, which continues today, fit the dreams of the Newborn Red Nobles?

Let me now step back and try to explain events using a poem by Mao Zedong, written in 1973, three years before his death.

The poem deals with the political system in the Qin dynasty, the first ruling dynasty of Imperial China.

Qin Shi Huang, the first emperor and the strongest, relied on no one for his power but his own iron will. He did not rely on the authority of Confucianism. Why should he rely on someone who was only a thinker and educator? Qin did not need this kind of person to maintain order, and neither did Mao.

In the poem Mao criticizes an article by Guo written in 1948 while Chiang Kai-shek was still in power. Guo compared Chiang Kai-shek to the autocratic Emperor Qin, and praised Confucius.

Here is Mao's poem (translated by me):

*Take a break from your scolding of Qin, the
   first Emperor,*
*There are other opinions about 'book burnings
   and elite killings.'*
*The Grand Dragon has died but his spirit
   remains still over the world*
*Confucius, enjoys a lofty reputation, but is
   nothing at all but nonsense, bullshit.*
*Qin System was employed by each of the
   dynasties in China,*
Your criticism of him is not good.
*Read 'On Feudalism' again and again from
   Liu in the Tang dynasty*
*Don't go back from him [Liu] to Wen, the King
   of Chun Qiu era.*

After the Qin Dynasty in 200 BC, the 'Qin system of Rule' has lived on through the iron hand and centralized rule.

The emperors in China, of every dynasty, have continued to exploit the ordinary people just as Qin did. They levied farm rents and taxes, they employed forced labour, and they used whatever means of exploitation they wanted.

At the same time, to discourage dissent, each emperor promoted his own brand of Confucianism to promote deference and maintain social order.

For two thousand years, this kind of imperial control remained the status quo.

In the early twentieth century, when the last Emperor of the Qing Dynasty (1644–1911) was kicked out of the palace, did this usher in a new spirit of change for China?

Okay, Chinese men cut their braids, wore Western suits, started talking about having a democracy, a republic and a Constitution. The whole society bustled with excitement.

But the country's old centralized political and legal system remained unchanged; stubbornly stuck in the spirit world of the Chinese people, in the heart of both the rulers and the ruled alike. But now China had modern-day emperors – President, Chairman, General Secretary – one after the other.

It remained under Chiang Kai-shek's Nationalist government; it remained under Mao Zedong's People's government. The ghost of Qin – or 'Grandpa Dragon', as Chairman Mao used to call him – continued to haunt the Chinese political psyche.

Chiang Kai-shek was unlucky, because not even one day of his era, from 1927 to 1949, was a day of peace (warlord war, anti-Japanese war, civil war).

Then Mao seized power. In the following twenty-seven years before he died, China should have enjoyed many years of peace . . . but what actually happened?

Declassified documents, released in September 2010, reveal that in 1950 Stalin made a deal with Mao: 'I'll manage the European side of things, you'll be responsible for the Asian side,' Stalin is recorded as having said.

I don't know how many high-ranking officials in the inner CCP knew about that deal.

But we know that after the deal was made, Mao began to see himself as a leader, not only of China, but of the oppressed peoples of the East. It became his dream, his 'responsibility,' to reach beyond China's border.

This explains why, in the early 1950s, he resisted the advice of nearly all of his comrades and made China 'pull chestnuts out of the fire'* for Stalin, and act as 'the cat's paw' on his behalf by becoming involved in the Korean War.

This also explains why three generations of Chinese people (born from 1910 to the 1970s) were forced to subsist at a miserable, survival standard of living, especially those farmers who lost their land and became 'commune members' but in fact, were no more than farm slaves.

None of the Communist elite, from the top level in Beijing to the grassroots level in

* In other words, let others do the work when the job gets too difficult. It is based on the fable of the monkey that asks its companion, the cat, to help remove the roasting chestnuts from the fire with its paw. While the cat did so, the monkey ate the roasted chestnuts, leaving the cat with nothing but burnt paws.

villages, dared to question Mao, their 'emperor,' about his strategies, no matter how absurd they were – when he launched endless 'movements': the land reform, the capitalist reform, the Great Leap Forward, the anti-rightists' attack on intellectuals, and the internal struggle. All of these movements led to famine in the 1960s – a time of peace, causing more than 30 million people to die from mass starvation.

And so, by the 1970s, China's economic situation and the livelihoods of its people were on the brink of collapse.

And then Mao Zedong died. The faction that inherited his spirit and doctrine – the Gang of Four – were purged.

Once again, the people raised the cheer of 'Liberation!' But it is impossible to purge the political culture of two thousand years of centralized imperial rule from the minds of the people at the drop of a hat.

Then Deng Xiaoping came to power and imposed the traditional way of centralized rule to stop the chaos that had emerged under Mao.

As he said to Jiang Zemin, 'Chairman Mao

had the final say when he was alive, and now that he's gone, I have the last word. When you are in this position and have the final say, I can rest my heart.'

But what about the Chinese Communists – former rebels and idealists – who followed Mao in the agony of the Long March; who followed him into the so-called 'Anti-Japanese War'; who held the red flag with tears in their eyes when the People's Republic of China was established in 1949.

What kind of 'Communist fighters' should they now be? And what sort of ideals should they follow now?

China began to 'reform and to open its door to the world'. And the whole world responded to this news with a warm welcome. This gave China a great opportunity to integrate with the outside world and to acknowledge shared universal values – just as the Eastern European countries, and even Russia and Vietnam, had done.

I must say that in the 1980s, this is generally what the situation looked like. Unfortunately, when students went to Tiananmen Square in 1989, the bloody suppression changed China. It completely changed the trajectory of China's modernization, with the most visible change being in the 'soul' of China's political elite.

Before Tiananmen Square, during the Mao era and the 'reform and opening up' of the 1980s, members of Mao's political elite never challenged their emperor's power. They were heavily influenced by an idealistic belief in socialism, public ownership, a planned economy and a commitment to 'Serve the People'.

After 'Deng Xiaoping's Southern tour' in 1992, when he declared that 'getting rich is glorious' the Communists, officials, their families and the ones around them, began to build a 'power-linked capitalist system'.

To achieve this goal, the new Red China established a unique operational system. It completely marginalized the idealistic early Communist supporters, replacing them with Red Technocrats pursuing practical interests.

This capitalist system, however, was not a market-driven economy. Instead, the drivers were the ones who spent the state's assets as if those public assets were their own.

This soulless system has been the state of affairs for nearly twenty years.

Following Deng's declaration of 'development as the top priority' came Jiang Zemin's 'quietly making a big fortune.' Under the banner of 'representing the interests of the people' the Red Technocrats became more and more powerful.

At first, they snatched money through special policies and channels, such as the foreign currency exchange and shortages of materials. Then they turned to making money through the sale of weapons and, thanks to the privilege of government access, the operation of projects.

These money-making opportunities are no longer enough. The Red Technocrats have since moved into banking and stock market manipulation, as well as real estate and land development – which has become the most important source of income for local government officials.

President Hu Jintao and Premier Wen Jiabao are continuing along the same path laid out by

Deng. Their central strategy is for 'the national enterprises to advance, while the private companies give way.'

In eight years of their administration, governments at all levels have enacted administrative measures to monopolize highly profitable enterprises – in shipping, railways, electricity, energy, telecommunications, and in other major sectors.

The CEOs of China's state-owned enterprises have personally gained through a 'two accounts' policy that has ensured the protection of their monopolies by the government while they pocket the profits. The salary of a general manager in a state-run enterprise might be a hundred times more than that of a worker. These bosses make up the backbone of a powerful and corrupt privileged class.

When Hu stresses that the government is 'maintaining (social) stability' and 'avoiding self-inflicted setbacks,' what this really means is that they are 'busy making money by exploiting the people, so please do not disturb.'

*

What kinds of lives have the Chinese people led for the past sixty years under single-party state rule?

There is a simple saying from netizens that sums up the sixty-year period of the Red Empire of the People's Republic of China – from Mao (Zedong) to Deng (Xiaoping) to Jiang (Zemin) and then to Hu (Jintao).

This saying has spread far and wide on the Internet and it goes like this:

> *In the name of revolution, they justified killing.*
> *In the name of the people, they justified*
>   *nationalization.*
> *In the name of reform, they divided the spoils*
>   *of the nation.*
> *In the name of harmony, everyone must now*
>   *shut their mouths.*

But not everyone shuts up. Liu Xiaobo, China's first Nobel Peace Prize winner, refused to shut up and the ruling elite jailed him for eleven years. Ai Weiwei, China's most bold and daring artist, and the poet Tan Zuoren also refused to shut up when they tried to expose how many children were killed by poorly constructed 'tofu' schools that crumbled in the May 2008 earthquake. They were beaten and put into jail for voicing their conscience.

Today, in China, Communist elite interest groups have emerged. In terms of different sectors, these groups include people from all walks of life: the Party, government and military; the fields of science and technology, education and culture, medicine, healthcare . . . everything.

In terms of power, these groups include all levels from top to bottom: from the Communist Party Politburo Standing Committee, to the province, city, county, township, village, down to the lowest level.

In this group, everyone is trying to exploit their power for personal gain and for the benefit of their family.

Their approach is to 'make the cake bigger' so they can 'cut the cake' and share the pieces by ignoring either the legal system or civilian oversight.

How do they make the cake bigger? By taking the easiest, safest, and fastest way, of course: by exploiting 'the vulnerable groups.'

As Qin Hui, the eminent professor of history from Qinghua University says: the secret of China's 'rise' lies in 'the advantage of low human rights' – allowing for the exploitation of the rights of provincial workers, natural resources, and the environment.

In China, we have no independent trade unions, farmers' unions, no chambers of commerce or industry associations – only countless silent workers who have no sense of rights and no channels of complaint.

We have no independent media or independent academic research. All avenues of communication – television, radio, newspapers, publishing houses, research institutes and universities – are either mouthpieces of the government or subject to the party's control and censure.

We have no independent or registered human rights or environmental NGOs, or independent foundations. Those public interest researchers and lawyers who try to be watchdogs and uphold the Chinese Constitution are themselves watched and suppressed when they try to contribute to the peaceful transition of China to a country of laws.

We also have no meaningful protections for the environment.

According to the Constitution, China's land, rivers, forests, and mineral resources are all state-owned.

In practice, this means owned by state officials. Any official who puts his hands on our resources can own it. Land grabs have become the primary means for officials, at all levels, to get rich.

Since 1990, over the past twenty years, this system and China's 'rise' has led to the disastrous destruction of China's resources and environment.

- 80 per cent of the rivers and lakes are drying up;
- 60 per cent of the water in seven major river systems is unsuitable for human contact;

- One third of the land is contaminated by acid rain;
- Two thirds of the grassland has become desertified and most of the forest is gone;
- Water systems and soil have been severely polluted by fertilizers and pesticides, with 40 per cent of arable land degraded.

China has become the world's factory and the world's dumping ground as well. Of the world's twenty most polluted cities, sixteen are in China.

These are the costs to our resources and our environment. But what about the cost to the people of China and the nation, as a whole?

Nowadays, the career path most people hope to follow is that of government official, which is seen as the gateway to becoming rich. The traditional Chinese work ethic has disappeared.

In China today, with a belief in neither the traditional values nor the rule of law, money has become a be-all and end-all for almost everyone.

This has become the common view.

People play games with authority. At one level,

they act obsequiously; at another, they are envious.

They are driven by the sense that they, too, must become rich. To become rich, they will follow the example of their superiors, even if it means tyrannizing the weak and plundering public resources.

In today's Qin Policy, the Chinese know that they are free to do anything, as long as they remain silent about politics. They can chase money – no matter how immoral. The rulers get the lion's share of the spoils and hand out small morsels to those who follow the rules of the game, and know enough not to challenge them.

The Qin System, as Mao said, has been employed by every dynasty in power. This system still continues but in the hands of today's communists, it is executed with great efficiency and skill.

But, what about the environment, is it sustainable?

Friedrich Hayek once said that 'a tyrannical government without any restriction means only war and enslavement.'

Today, because we have an authoritarian system in China, our resources, our environment and the welfare of the people are not secure.

China's great 'rise' is no rise at all. It has meant destruction for the country's rivers, land, forests, the children of China and the nation itself.

What can we, as residents, citizens and netizens in China, do in this kind of political environment?

The only way, I think, is to tell the truth about the costs of the 'rise'; to act not only as a netizen but as a true citizen with the basic right to free speech, assembly and public oversight of government; to fight the twenty-first-century's Qin-style dictatorship; and to insist on fighting for our constitution, and not for rebellion or revolution.

Though the battle will be long, we will not give up.

*The Three Gorges Project**

A Symbol of Uncontrolled Development
in the Late Twentieth Century

*'Water benefits all things generously and without
strife. It dwells in the lowly places that men disdain.
Thus it comes near to the Dao.'*

– Laozi

The opening of my country to the outside world
has been the most important development in

*This essay was first published in *The River Dragon has
Come!* (1998), a follow-up to Dai Qing's ground-breaking
polemic *Yangtze! Yangtze!* (1989). The title echoes the cry
of a witness to the 1975 collapse of the Banqiao and Shi-
mantan Dams in 1975 which killed thousands.

twentieth-century China. The two major consequences of this 'opening' have been the birth, development, and dominance of the communist/socialist system, and the influx of modem science and technology. We Chinese are repeatedly told that both the communist system and the ascendency of science and technology fit China's historical conditions of economic underdevelopment, foreign domination, and political autocracy. But rather than 'fit' our national conditions, these systems have dominated and distorted our lives. As the old Chinese adage says: 'Things will develop in the opposite direction when they become extreme' *(wuji bifan)*. This is the case with our current socialist regime and its blind faith that engineers and technical fixes can solve all problems. The result of all this is uncontrolled development, and there is no better symbol of uncontrolled development than the Three Gorges dam.

'Uncontrolled' *(bujia jiezhi)* and 'out of control' *(shiqu kongzhi)* are similar terms which actually have different implications: The first – uncontrolled – is subjective and describes someone who consciously fails to control his/her behaviour. The second – out of control – is more objective and describes how someone's behaviour can cause things to spin out of control.

The Three Gorges project has been meticulously planned and controlled from its original design to its final construction. But the people who have been doing this planning have failed to understand key Chinese concepts such as self-restraint and the control of brazen arrogance. In Chinese antiquity, a sense of self-restraint was paramount; as the ancient Daoist philosopher Laozi said: 'To know one's limits is to be invincible' *(zhizhi keyi budai)*. But a couple of centuries after the advent of the Industrial Revolution, this ancient wisdom lost its appeal and has only been recalled in the last fifty years. This conscious failure by China's leaders to 'control' their behaviour; that is, to respect and follow ancient wisdom, is what makes the Three Gorges dam a symbol of uncontrolled development. The sad irony is that although every aspect of the Three Gorges dam's construction has been thoroughly planned by scientists, engineers, and officials, if it is completed and goes into operation, we will quickly learn that we are unable to control its effects on the environment, and on society.

The Three Gorges dam will be the largest dam ever built. Its wall of concrete, reaching 185 meters into the air and stretching almost two kilometers across, will create a 600-kilometer-long reservoir.

The dam will require technology of unprecedented sophistication and complexity: it will include twenty-six 680 MW turbines; twin five-stage lock systems, and the world's highest vertical shiplift.

The project will also cause some of the most egregious environmental and social effects ever: it will flood 30,000 hectares of prime agricultural land in a country where land is the most valuable resource; it will cause the forcible resettlement of upward of 1.9 million people; it will forever destroy countless cultural antiquities and historical sites; and it will further threaten many endangered species, some already facing extinction.*

---

* Some of the most seriously endangered include the white-fin dolphin (whose population now numbers less

But perhaps the most astounding fact of all is that although the project has attracted the interest of the world's businesses and the ire of its environmentalists, it has faced very little opposition at home. The National People's Congress (NPC) approved the project in April 1992, but since then very little has been said or written in opposition to the dam that will disrupt the lives of so many and damage such great swathes of our territory.*

Everyone knows that China is facing an energy

---

than one hundred and is on the verge of extinction), Chinese sturgeon, Yangtze sturgeon, yanzhi fish, white dolphin, and river sturgeon.

* NPC approval of the dam came in April 1992, by a vote of 1,767 in favour, 177 opposed, and 644 abstentions, an unusual display of public opposition in the generally rubber-stamp body. The vote was in favour of a resolution to authorize construction of the dam and was conditioned on a promise from the Three Gorges Project Development Corporation to resubmit more precise construction schedules for future approval. See Dai Qing, *Yangtze! Yangtze!*

Three Gorges Dam Specifications

Dam crest	185 m
Dam length	2,000 m
Reservoir Functions	
Normal pool level	175 m
Flood control level	145 m
Total storage capacity	39.3 billion m³
Flood control storage	22.1 billion m³
Navigation	Reservoir level raised by 10–100 m to allow 10,000-ton ships to Chongqing
Power Generation	
Installed capacity	17,680 MW
Unit capacity	26 units, 680 MW/unit
Inundation	
Land	632 km-long, 19 cities, 326 towns
Arable land	430,000 mu [30,000 hectares]
Population	1,130,000 people

Note: Figures for land inundated and people moved are government estimates and are questioned by dam opponents.

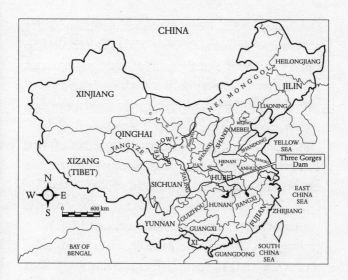

shortage,* that our transportation systems are
congested, and that we suffer frequent floods.
The country has only recently emerged from

* In 1994, China generated 926 billion kilowatt-hours of
electricity, 19 per cent of which came from hydropower.
Demand for electricity, which has been substantially
underpriced in China's centrally planned economy, is
expected to grow at an annual rate of at least 6 per cent
through the year 2000. See *China Statistical Yearbook, 1996*,
pp. 203–207.

the chaos of the Cultural Revolution (1966–76) and, with the pursuit since 1978 of a new, more open economic policy, increased foreign trade, and dramatic reforms in agriculture and commerce we have finally begun to experience some remarkable economic gains. Why then, just when the country seemed to have a bit of money to spare, was this mammoth project proposed; especially when there were smaller and more viable options to meet our energy, transportation, and flood control needs?

The best alternatives involve building smaller dams on the Yangtze River's tributaries. But alternatives were never seriously considered by the top leadership. Why? Because China is in the midst of a phase of 'uncontrolled' development where a sense of moderation and restraint are completely absent. This lack of control is evident at every level of planning for the Three Gorges project: From the 'red specialists'' faith in technology, to the closed decision making of autocratic leaders, and the complete disregard for the environmental effects of the project on the river valley and its residents.

The Power of the Red Specialists

红色专家

In China the so-called red specialists (*hongse zhuanjia*) consider themselves infallible even though the history of the People's Republic is littered with grandiose technological and economic projects gone wrong, often at enormous costs to the treasury and to human life.[1] With regard to the Three Gorges dam, this sense of infallibility manifests itself in a number of ways. For instance, the red specialists arrogantly claim that they have the technical ability and capacity to build the world's largest dam, turbines, and shiplift. But what they fail to consider is that the use of this technology does not make hydrological and environmental sense. Meeting the difficult technical challenges posed by the project should never take the place of sound scientific decision making. Decisions based only on what is technically possible may eventually succeed in building the dam and turbines, but they are unlikely to solve the pressing hydrological, environmental, and human problems which the dam will undoubtedly cause. This point was raised as early as the 1930s by Professor Huang Wanli. But unfortunately, the opinions of such venerable sages have had virtually no impact on policy that is driven by visions of technological grandiosity.

Even if the Three Gorges project is completed at the appointed hour, the long-term upheaval and damage caused by the resettlement of upward of 1.9 million people and the destruction of treasured cultural relics will be difficult, if not impossible, to reverse. The havoc created by the vast resettlement scheme will not only carry an immense price tag, but will also forever damage the spiritual and psychological health of the relocatees. The dam is not just about the loss of beautiful tourist landscapes, but about the damage the nation will do to itself through the patent disregard and ignorance of its spiritual wealth.

The 'red specialists' have never managed to grasp the concepts of fundamental order and balance in the relationship between humankind and nature. At every turn – from its preference for a planned economy with a focus on iron and steel production, to its promotion of grain production, population growth, and large-scale dam construction* – the Chinese leadership has made decisions which run counter to the Chinese philosophical concepts of maintaining order and balance between humankind and nature. Not surprisingly, each of these decisions has caused immense damage to the country's environment and natural resources. For political reasons, however, those scholars and intellectuals who are in touch with this philosophical tradition have had very little opportunity to speak up. With the promotion of a new market economy since 1978, profit once again comes first in the minds

* A reference to policies in the 1950s promoted by Mao Zedong over the objections of some scientists, agronomists, and hydrologists that led to converting almost all available land to grain production in order for each region to achieve agricultural self-reliance, that opposed population control on the grounds that more people meant more power for the 'new China,' and that led to a massive dam-building campaign during the Great Leap Forward (1958–60).

of China's leaders, and all they think about is plundering nature rather than respecting and conserving it and maintaining the balance.

Adding to the problem is the fact that so many of these specialists make decisions based on blind self-interest, or on the narrow interests of their bureaucratic bailiwicks.

Reckless actions by specialists and bureaucrats who possess narrow scientific and technical skills can be very frightening indeed. Such people plan things in very meticulous ways to fit their own personal interests and work only to advance the goals of their respective bailiwicks. They could care less about the national interest and the fate of the nation.

Autocracy and Closed-Door Decision Making

Throughout its history, China has been ruled by an autocratic system. In the distant past, everything was done in the name of the emperor. After the 1912 Republican revolution, it was done in the name of the people's revolution. And since 1949, during the reigns of Mao Zedong and Deng Xiaoping, it has been done in the name of communism and socialism. Autocracy is still considered an acceptable form of government in some areas and under some circumstances either because there is no alternative system, or because it is believed to be appropriate at a certain stage of a nation's development. Nevertheless, autocratic governments are on the wane.

The Three Gorges project has both benefited from China's autocratic history and helped strengthen it. Those promoting the dam, from the 1950s to today, have all been masters of political gamesmanship, constantly referring to 'Chairman Mao's desire' (*Mao zhuxide xinyuan*) and 'Deng Xiaoping's support and concern' (*Deng Xiaopingde zhichi he guanxin*) for the project. By invoking the support of the country's autocratic leaders, the dam was made virtually unassailable.*

* In 1953, Mao first expressed interest in the Three Gorges dam and insisted on building a single large dam, instead

When the project did run into resistance, the dam-supporters used nationalistic bluster to reinforce their position. Nationalism is an inherently parochial, irrational, and extremely destructive force that ultimately runs counter to the interests of human development. It should only be called on in extreme circumstances, such as in resisting foreign invasion, and not otherwise used to stir passions and excitement.

Although private companies and other ostensibly private organizations have been established to assist in the construction of the dam, the project has relied on government financing since its inception. Given that China is trying to move in the direction of a market

of a series of smaller ones on the Yangtze's tributaries, something that had been proposed by the hydrologist Lin Yishan. Mao even suggested that he might resign as chairman of the Chinese Communist Party to assist in the project design which was eventually overseen by Zhou Enlai. *Mao Zedong zai Hubei* (Mao Zedong in Hubei Province) (Wuhan: Hubei People's Publishing House, 1993), pp. 95–100.

economy, the decision to build a large project such as the Three Gorges dam solely on the basis of the leadership's will can only have a negative impact on the transition.

Government munificence has come in many forms: direct allocations by the state; the transfer of revenues from the Gezhouba dam; and increases in national electricity rates. The government has also 'recommended' that some profitable large enterprises 'assist their counterparts' through donations to the Three Gorges project. This sort of action strengthens and supports the central planning apparatus in the economy and works to stifle independent thought and competition.

Because local leaders are centrally appointed under China's autocratic system, they do not dare strive for a fair deal for their local constituencies. The people of Chongqing, Sichuan (who will receive few if any benefits from the dam and may suffer many of its negative effects), have condemned their leaders for selling out Sichuan's interests. In 1996, Chongqing Municipality was granted province-level status under the direct authority of the central government, thereby separating it from Sichuan Province, whose leaders have generally not supported the dam.[2] Even more significant is that, in 1989, amid strong opposition to the dam, the State Council decided to postpone consideration of the project. But in the political atmosphere following the Tiananmen Square massacre, all opposition to the project in the government was crushed, and 'senior leading cadres' used their political weight in the traditional style of autocratic politics to ignore legal procedures and ensure that the project went forward.[3]

Subsequently, when the Three Gorges project was awaiting approval from the NPC, the national press was mobilized to write only positive reports about it. Meanwhile, even before the NPC convened for its vote, the chair made it clear that its approval was not in question.[4] During the course of the session itself, the microphones on the floor of the NPC were turned off to prevent the dam opponents among the delegates from voicing their views and generating collective opposition.

China's autocratic leaders have used the most undemocratic procedures imaginable to push the project forward. I don't think for a moment that China's modernization can be achieved overnight, but the government and the people should break with the traditional autocratic system and make a conscious effort to gradually begin the transition to a more open system in order to bring about a fundamental transformation in China's political culture. Instead, supporters of the Three Gorges project continue their efforts to consolidate power and support the old system by whatever means necessary in order to ensure that the construction goes forward.

The Effects of Uncontrolled Development on the Environment

Even if construction of the Three Gorges dam is completed as planned in 2013, its ability to generate electricity depends on avoiding a massive build-up of sediment behind the dam. Because of sedimentation, the Three Gate Gorge

Construction Phases

Phase	Year	Construction stage	Water level
Preparation 1993			
First 1994–97	1994	Excavation of base begins Project inaugurated	
	1995	Pouring of reinforced concrete begins	
	1997	River blocked and diverted	
Second 1998–2003	2003	Electricity generation begins	135 m
Third 2004–2013	2007		156 m
	2009	Electricity-generating system completed	
	2013	Normal operation	175 m

dam (*Sanmenxia*) on the Yellow River has induced floods in the river's upper reaches and led to the resettlement of over 400,000 people. It now produces less than one third of the power that was promised, its turbines are damaged by sediment, and it will not be able to fulfill its flood-control function until another massive dam, the Xiaolangdi, is built downriver.

After the Three Gorges dam, the Xiaolangdi dam is the second largest such project in China. Slated for completion in 2002, it will cost U.S.$3 billion and will involve the relocation of over 400,000 farmers.

The Three Gorges dam will face similar sediment-related problems. Even if the dam does generate the promised electricity, most of it will go to serve southern and eastern China. Sichuan Province will be unable to develop its own regional electrical supply because all of the money available for electricity generation is tied up in the Three Gorges project. The province will reap few benefits from the dam, but will bear many of its costs, especially the loss of land and the burden of resettlement.

The primary purpose of the Three Gorges dam is flood control, and it has been designed to contain a once-in-one-thousand-year flood. But no single dam could ever contain such a flood on the

Yangtze River. Unable to contain massive floods, the Three Gorges dam provides, conversely, an excessive and unnecessary level of protection from the smaller floods which frequent the Yangtze. Even at its peak, the 1981 flood in Sichuan Province never reached the cities of Yichang or Wuhan.[5]

From the beginning of the dam project, Huang Wanli has consistently warned the leadership against creating a situation similar to the 'Railroad Protection Movement in Sichuan', which, he noted, 'led to the 1912 Republican revolution.' The Railroad Protection Movement was an immediate cause of the 1911 revolution that overthrew the Qing (Manchu) dynasty (1644 – 1911). The movement was centered in Sichuan where local merchants resisted the central government's railroad nationalization plan because it entailed foreign loans, fostered official corruption, and led to the imposition of commercial taxes to finance the entire scheme.

That revolution, we now know, turned out to be enormously destructive. People in China and throughout the world sincerely hope that the country's transformation and modernization can be carried out smoothly, but the Three Gorges project runs counter to this hope because, in its name, the government has suppressed free speech and strengthened its power at the expense of the provinces and the people. The project is encouraging corrupt economic practices in enterprises and in the government and will lead to a vast waste of resources, all the while destroying the environment and violating the rights of the people.

We are fortunate that we live in an open world, for the effects of the Three Gorges project transcend national boundaries. If the project is to be supported financially by multinational organizations, then it cannot avoid the scrutiny of the outside world.

The human race has readily demonstrated its capacity to destroy the environment, and we do not yet know how to control our desires and greed. So what should we do when such an uncontrolled project is being carried out under the watchful eye of the Chinese public? I know that other countries subject their hydropower projects to public scrutiny with success. But how can the Chinese people struggle for the same assurances in the case of the disastrous Three Gorges dam?

Notes

1. Grandiose engineering and energy projects have also been criticized in the former Soviet Union. See Grigori Medvedev, *No Breathing Room: The Aftermath of Chernobyl*, trans. Evelyn Rossiter (New York: Basic Books, 1993).
2. Province-level conflicts and divergent interests over the dam are analysed in Kenneth Lieberthal and Michel Oksenberg, *Policy Making in China: Leaders, Structures, and Processes* (Princeton: Princeton University Press, 1988). See also, Epilogue.
3. This information was contained in a summary of the Lunar New Year Forum published in *Xinhua Monthly,* which covers domestic developments. [Note: Many sources provided in the original text are incomplete, eds.]

4. According to a participant at the meeting who wishes to remain anonymous.

5. Evidence also exists that Chinese government officials grossly exaggerated the severity of floods to justify construction of the Three Gorges dam. See Simon Winchester, *The River at the Center of the World: A Journey Up the Yangtze, and Back in Chinese Time* (New York: Henry Holt and Company, 1996), pp. 220–230.

The race to salvage the
*Three Gorges treasure trove**

Few people realize that two main civilization systems developed in parallel in ancient China: the Yellow River civilization of the Han culture, and the Yangtze River civilization of the Ba and Chu cultures.

Chinese archeological work began in earnest in the 1920s. And until the 1970s, archeologists concentrated their attention on the Yellow River valley. Their neglect of the Ba/Chu civilization along the Yangtze River did not mean they were unaware of its importance. It was just that the Han sites were located in the heart of the Yellow River – in the middle reaches of the river and on

* A talk delivered at the Cleveland Museum of Art on 27 March 2002.

the plains of the river valley. These sites were so much easier to reach and to excavate than the more remote, mountainous Yangtze River sites of the Ba and Chu people.

Since the middle of the 1980s, when the central government revived an old plan to build the huge Three Gorges dam, many Chinese archeologists have been terribly worried about the region's archeological treasures. Numerous historic sites and cultural relics are to be submerged forever under the dam's reservoir. The experts are well aware of the archeological treasure trove in the Three Gorges area, though no accurate record has ever been made of everything that's there. And they're also painfully aware that losing these ancient treasures before they are properly surveyed and excavated will represent a major loss to the study of world civilizations in general, and to archeological work in China in particular.

More than 60 archeological sites and ruins containing fossils and other evidence of extinct life forms and of human activity dating back to the Old Stone Age have been identified in the Three Gorges area. Fourteen of these sites have remained completely untouched, including an open-air site discovered in the town of Gaojia in Fengdu county, where stone tools were produced in the Old Stone Age.

A large number of stone tools, such as choppers and scrapers characteristic of south China, have been found at these sites. These show that the Three Gorges area contains invaluable information pertaining to the two civilizations that developed in north and south China. This important evidence could help determine the dividing line between those two Paleolithic cultures.

More than eighty archeological sites and ruins dating back to the New Stone Age have also been found. These could be the key to identifying the dividing line between the cultural systems that developed in east and in west China – specifically, the culture of the Jianghan Plain to the east, and the one to the west, in the Three Gorges area and beyond, in the Sichuan basin along the Yangtze River.

More than a hundred historic sites and tombs belonging to the Ba people have been identified in the Three Gorges area. These Ba sites functioned as political, economic and cultural centres from as early as the Xia dynasty, which began in the twenty-first century BC, up to the Qin dynasty, which ended in 206 BC. Proper archeological investigation of this area would help us gain an understanding of the intriguing Ba culture, about which little is known.

Studying other archeological sites and tombs in the Three Gorges area would allow us to figure out how the Chu culture and the Qin culture developed and spread in the region. The influence of the Chu culture had reached Xiling Gorge, around the Zigui area, in the mid or late Western Zhou Dynasty, which began in the eleventh century BC. The Qin people conquered the Chu in 278 BC, occupying the Chu capital, Yindu, near present-day Jingzhou in Hubei Province, and ruling over the whole Three Gorges area.

The Zhongyuan culture, which formed part of the Yellow River civilization, was situated mostly in Henan province. To understand the Zhongyuan culture and how it interacted and gradually blended with that of the Ba people, archeologists need to properly excavate and study 470 sites and tombs, including the two ancient county seats of Yufu and Juren. The dates of all of these sites range from the Han dynasty, which began in 202 BC, to the Six Dynasties period, which ended in 589 AD.

Six sites contain ancient inscriptions carved in stone that record dry-season water levels, while 10 other sites record the flood-season water levels. These sites, containing precious hydraulic information about the Yangtze River, date from the Song dynasty, which began in the tenth century AD. One of the best known of these sites is Baiheliang (White Crane Ridge), which may be the world's oldest hydraulic monitoring station.

Other things that will disappear include stone carvings made in the Eastern Han dynasty, which began in 25 BC; and dozens of Buddhas and stone tablets carved with poems and prose dating back to the culturally rich Tang dynasty of the seventh to tenth centuries. These are not only precious works of art; they also contain irreplaceable information about the region's history.

Hundreds of magnificent structures dating from the Ming and Qing dynasties – from the fourteenth to the early twentieth centuries – will be lost. These include temples, houses and bridges, set against beautiful natural landscapes and providing a wealth of information about China's traditional cultures. A host of relics belonging to the Tujia and other ethnic minorities help trace 4,000 years of cultural history in the Three Gorges area.

Other losses include some of the world's most important relics related to ancient transportation methods – the plank roads carved into the sides of the steep mountains beside the Yangtze, and the towpaths built along its banks.

The early rulers of modern China were military men, and their successors in the 1980s were engineers. It is little wonder that no matter which of these leaders was in power, no care or concern was shown for preserving China's human heritage or its natural resources and environment. In the leaders' minds, industrialization – and later the stock market – were far more important.

No sociologists, anthropologists or archeologists were invited to take part in the feasibility studies for the Three Gorges project. In the initial budget for the dam, one-third of the funds were earmarked for construction, one-third for electricity transmission systems and one-third for resettlement costs. Not a single penny was set aside to help salvage historical relics.

Since the project was given the go-ahead in 1992, many Chinese scholars, writers, artists and archeologists – including the discoverer of Peking Man – have appealed to the central government to devote funds to this salvage operation. Finally, under growing pressure at home and abroad, the Three Gorges project authorities decided to take about US$60 million out of the US$5-billion budget earmarked for resettlement and put it toward the cultural rescue operation. These funds represent just one per cent of the resettlement budget.

In these circumstances, about thirty Chinese universities and research institutes have had to raise their own funds to send archeological teams to the reservoir area. It is thanks to their work that we now have a rough idea of what's there and what will be lost. I'm told that these archeological workers are often staying in shabby hotels costing little more than a dollar a day. And because of their desperate lack of funds, about all that these researchers can do is survey what will be lost, rather than actually excavate sites or undertake much in the way of salvage and preservation.

More importantly, government and project authorities have paid little attention to the salvage operation. It seems extremely difficult for them to take these sorts of things seriously. As a result, the situation has handed thieves and smugglers a golden opportunity to steal the treasures and sell them on the black market.

It came as a real shock when a priceless bronze 'money tree' unearthed from the Three Gorges area and dating back 2,000 years to the Han dynasty was sold for US$4 million in New York in 1996. Premier Zhu Rongji, who is in charge of the Three Gorges Project Construction

Committee, was outraged when he heard the news. After his personal intervention, an additional US$375 million was earmarked for the salvage work.

The Three Gorges reservoir is due to be filled in June 2003. This socialist megaproject will uproot two million people, submerge 48,000 acres of good-quality farmland, endanger some precious animal and plant species, and plunge many priceless historical relics permanently under water.

Under Mao Zedong, the central ideology in China was class struggle. Under Deng Xiaoping and Jiang Zemin, it became 'to get rich is glorious.' It is not hard to foresee the fate of the archeological relics in the Three Gorges area. You need only compare the attention that is paid to the different parts of Beijing: China's capital has magnificent new shopping centres – and poorly funded, dilapidated museums.

Who wants this dam anyway?*

Given that there are so many problems with the Three Gorges dam, who on earth wants it to be built anyway? And how is it that the project could go forward under such circumstances?

The Three Gorges dam has been a pet project of several top Chinese leaders, including Mao Zedong and Deng Xiaoping, who saw the dam as a historical monument in which their greatness could be recorded by bending nature to their will.

The project has also handed several technocrats the opportunity to achieve personal goals such as political promotion or academic glory,

* An excerpt from a talk given at the Riversymposium conference in Brisbane, Australia on 5 September 2002.

followed by a comfortable retirement. These individuals don't really care about the people displaced, the rivers affected or the environment damaged by high dams and big reservoirs.

Li Peng, one of the most powerful backers of the Three Gorges project, was promoted from his position as a bureau director within a government ministry to the premier of the country in less than ten years.

Qian Zhengying, a former minister of water resources and now seventy-nine years old, not only became a senior member of the Chinese Academy of Sciences – a great honour for an official who quit college to become a revolutionary – but was also given a high-ranking position (vice-chairperson of the Chinese People's Political Consultative Committee) after her retirement. Though officially she is in charge of cultural, educational and public health affairs within the CPPCC, she is still a powerful figure in China's water resources and electric power industry.

Professor Zhang Guangdou, now ninety years old, holds two of the highest academic honours in China: He is a senior member of both the Chinese Academy of Sciences and Chinese Academy of Engineering. Because of his status within China's scientific community, he has helped to persuade younger scientists and engineers to follow the dictates of the Party and to defend government policies on this and other large dams.

None of the consequences of the Three Gorges dam are likely to affect any of these individuals or their families. No matter how severely the Yangtze River is damaged, and no matter how miserable local people's lives become as a result, these high-level backers of the dam will be quite unscathed.

Other groups of people also have an interest in seeing this disastrous project go ahead: hydropower engineers who lack even a basic knowledge of other scientific fields, including social science; corrupt officials in love with power and wealth; companies inside and outside China eager to profit from project-related contracts; foreign politicians and bankers who seek to obtain political and economic advantage in exchange for their support for the dam; and, within China, a group of radical nationalists keen to have the world's biggest dam in their country.

China's undemocratic political system and semi-capitalist economic environment have combined to allow the Three Gorges dam to be built. But the project is bringing irretrievable loss and harm to ordinary people and the environment.

And, in the absence of changes to China's political and economic systems, the Three Gorges dam won't be the last such megaproject in the country; the Yangtze won't be the last of China's rivers to be dealt such a blow; and the ordinary people who are being hurt by this project will not be the last group to suffer in this way. Encouraged by the Three Gorges example, arbitrary politicians and greedy businessmen are already extending their grasp to other rivers, such as the Heilong, Yalu, Leng, Nu, Min, Yaluzhangbu and the Lancang (or Mekong) rivers.

China has more dams than any other country

in the world. Just in terms of large dams, Australia has fewer than 500, while China has 22,000. I live in the 'most dammed' country in the world, and am part of the 'silenced majority.' We feel sorry for our rivers as we don't have the right to protect them, to express ourselves freely or to criticize the government openly, let alone the ability to monitor and curtail the government's actions.

Is 'keeping in step with the Party' good for the environment?*

Ever since the Chinese Communist Party came to power in 1949, Party members and the Chinese people as a whole have been endlessly admonished to 'keep in step with the Party.' The debate over building the Three Gorges Dam provides a good example of this.

In the early 1950s, very few people dared voice opposition to Chairman Mao Zedong's romantic idea of 'surprising the goddess of Wu Gorge by creating a huge man-made lake between the deep canyons.' Li Rui was an exception, but unfortunately he and his followers paid an extremely high price for their actions, which had failed to

* An article published in *Probe International* journal on 13 March 2002.

keep in step with the Party or with Chairman Mao himself.*

At that time, few people realized that building the Three Gorges dam would have such a harmful impact on the environment. Little was known in the 1950s about dam-and reservoir-induced environmental effects such as landslides, river-bank collapses, pollution, species extinction and so forth. But since the 1980s, we have been far more environmentally aware and now recognize that the environmental impacts of building the dam are too serious to be ignored.

China's three powerful authorities – the State Economic Commission, State Planning Commission and Chinese Academy of Sciences – conducted separate feasibility studies but all reached the same conclusion: that building the Three Gorges dam would be harmful to the environment.

* Li Rui, a former secretary to Mao and longtime opponent of the Three Gorges dam, was a vice-minister in the Ministry of Water Resources before being purged in 1959. He was rehabilitated in 1979.

In the feasibility study on the environmental impact of the Three Gorges project, scientists from the Chinese Academy of Sciences concluded that 'in terms of the environmental effects, the costs of building the Three Gorges project will greatly outweigh the potential benefits, and a high cost will be paid if the dam is built.'

In these circumstances, and in the face of growing opposition, the State Council announced in the spring of 1989 that the idea of building the Three Gorges dam would be put on hold for at least five years.

Unfortunately, after People's Liberation Army tanks rolled through Tiananmen Square on June 4, 1989, nobody inside China dared challenge the Party or the central government, and the different voices that had spoken out about the Three Gorges project were relentlessly silenced.

June 4 provided a golden opportunity for project proponents to put the dam back on the agenda and push it through as fast as they could – since the original decision had failed to keep in step with the Party. This climate handed the Ministry of Water Resources the opportunity to organize a new feasibility study, and it arrived at a conclusion completely opposite to the one drawn by the Chinese Academy of Sciences (albeit in the name of the CAS): 'In terms of environmental impact, the benefits of building the Three Gorges project will greatly outweigh any costs.'

Since then, non-government environmental groups in China have raised no criticisms of the dam, while government environmental protection departments have remained silent for years. Even several long-standing opponents of the project no longer spoke out. For instance, Lin Hua, former vice-director of the State Economic Commission and a well-known critic of the dam, said nothing, for 'the Party has made its decision.'

It was not until 1997 that a news report published by *China Environment Daily* (*Zhongguo huanjing bao*) prompted a strong reaction outside China. According to the report, the State Environmental Protection Administration (SEPA) admitted that building the Three Gorges dam would have a highly adverse impact on the environment. But soon, SEPA strenuously denied the report, and the reporter was said to have been severely punished – for being out of step with the Party.

However, not all government officials have been doing nothing while holding on to their taxpayer-supported positions; some are trying to help protect the environment. And so, even as the dam construction is at its peak, SEPA has published its 'Environmental monitoring bulletin of the Three Gorges Project 2000,' which includes some things that are not in step with the Party:

- Due to human activities, some rare birds in the Three Gorges reservoir area have been turned into visitors rather than residents.
- The area providing wood fuel is declining, and erosion has become a serious concern in the reservoir area because of the shortage of rural energy sources.
- Geological disasters, such as riverbank collapses and landslides, are increasing in the reservoir area, leading to growing economic losses.

- Pollution incidents are on the rise, caused by boats and by the garbage dumped directly into the river, seriously compromising water quality.
- No urgent and effective measures have been taken to deal with wastewater. Almost all polluted water is discharged untreated into the main channel of the Yangtze and its tributaries. Most garbage is washed into the river or heaped along its banks, creating a potential problem after the reservoir is filled.

Actually, none of this is new. Many scientists and engineers never stopped expressing similar or even stronger opinions in a variety of ways. It is also fair to say that government environmental protection authorities have held a more or less similar stance on these issues.

So while there is nothing new in this environmental bulletin, the fact of its publication indicates that these ideas are being allowed to be made public at this time.

The Yangtze and the Three Gorges*

Beginning in the mountains of Tibet, the Yangtze stretches 6,300 km through China before emptying into the East China Sea at Shanghai. On its way, it carves through mountain ranges and heads northeast before reaching the most impressive section of the river: the Qutang Gorge, Wuxia Gorge and Xiling Gorge, collectively known as the Three Gorges.

Once past the Three Gorges, the river widens and meanders through fertile plains and a number of major cities before entering the sea.

The Yangtze has always been essential for

* A speech delivered on a speaking tour in British Columbia, Canada in November 2010.

transport and agriculture in the central regions of China.

The total drainage area of the Yangtze is 1.8 million square kilometers, and is home to more than 400 million people. The Yangtze River valley is also China's agricultural and industrial heart, producing 70 per cent of its grain and 40 per cent of the total industrial output.

The Three Gorges region is considered one of the most important in China.

The Main Parameters of the Three Gorges Project

The parameters of the dam, as originally planned, have changed. The twenty-six turbines have become thirty-two – with another six small turbines installed underground. Originally, the Three Gorges Project (TGP) was to be completed in 2013 when the reservoir's water level was to reach 175 metres above sea level. But now, in the autumn of 2010, authorities have already raised the dam's reservoir to the 175-metre mark. The reason for this is simple: a higher water level means more electricity can be produced – and the sooner officials are able to raise the reservoir, the sooner they'll be able to generate this power.

Now, we have to stop here and ask: is damming the Three Gorges a success story or a looming disaster?

Authorities said the dam was necessary as it would produce power, provide flood control, improve navigation and benefit local citizens.

The government proudly declared that a rapidly modernizing China – with its 'socialist system with special Chinese characteristics' – had the financial and technical ability to build it.

The Project today

Before Mao's death in 1976, no one in China would dare say anything to challenge him – even if they were a top leader or one of his close comrades.

In the 1980s, China welcomed reform and opened itself to the world. Not just the country's new leaders, but society as a whole, began on a journey of self-examination – the air of enlightenment even spread to the intelligentsia.

A number of engineers, scientists and former high-ranking officials began to express, not just the opinions of the Party, but their own.

The most notable of these new voices – and one that captured the attention of the entire nation – was an Investigative Group from the Chinese People's Political Consultative Committee (CPPCC).

They were treated as allies and 'travellers in the same boat' by the Communist Party of China. But when it came to the fate of the Three Gorges and the Yangtze River, they decided to voice their dissent.

Their concerns and internal debates about the dam were published in the book, *Yangtze! Yangtze!* (1989).

Many of the problems they discussed have come to pass and many new, unforeseen problems have also emerged at Three Gorges.

Unfortunately, many of the questions raised by critics inside and outside China have never been answered.

Let's look at the benefits of the dam promised by officials:

Hydropower Generation

Power generation is the dam's main priority and the true aim of those who support it.

The TGP started producing power in 2003. To date, it has produced 390 billion TWH.

But consumers have yet to see the price of their electricity fall. In fact, electricity customers in China now have to pay a so-called 'construction fee' to support the project.

The generators installed at Three Gorges are 700MW, not 680MW. This is because, years ago, after Europe decided not to construct monster dams in their rivers, they then sold the technology for 700MW generators to Brazil. Brazil then sold it to China, which is the only country in the world with the guts, power and money to go forward damming mighty rivers. Now, as the most 'Powerful Hydro Nation' in the world, China has to get its money back on its dam-building prowess. This is why China has signed contracts to build major dams in Africa and Asia.

Flood Control

China is an ancient agricultural country. Emperors from the Ming and Qing dynasties, for example, to modern leaders – Sun Yat-sen, Chiang Kai-shek, Mao Zedong to Deng Xiaoping – have instinctively seen flood control as their primary task.

As such, the promoters of dams always scared these leaders with images of flood disasters as a way to appropriate the necessary funds.

In the 1910s, dam supporters began to approach Sun about damming the Yangtze, but their pleas were halted by the warlords' war. In the 1930s and 40s, Chiang Kai-shek was more occupied with the war against Japan. Still, he did send a delegation of interns to the US to learn how to dam the Yangtze.

Mao first wanted to build a major dam on the Yellow River, which eventually became the Sanmenxia dam. Then, he dreamed of building another dam, this time on the Yangtze with a "beautiful, smooth lake" in the Three Gorges region.

The dam builders argued about how high this dam should be, in order to store water and prevent floods.

But, even with the dam standing 185 meters high, would it be capable of completely stopping the threat of floods in Wuhan, the largest city below the dam and home to millions of people?

Year	Height of dam (m)
1940s	235
1950s	195
1980s	165
Now	185

In the summer of 1998, from June to August, the floodwater totalled 660 billion cubic meters – with 300 billion cubic meters coming from the upper reaches of the Yangtze and 360 billion cubic meters from tributaries below the dam. For the latter figure, the TGP is completely useless. Moreover, can a dam with a flood storage capacity of 22 billion cubic meters be expected to control such a devastating flood?

Zhu Rongji was appointed Premier in 1998. Just after his appointment he said: 'Everybody knows that I've been indifferent to the TGP, but now that I'm Premier, I have to deal with what has happened.'

What happened? The flood in 1998 is what happened!

He found, based on records, that it was not the biggest flood in history, but the water level was the highest on record. This is because sedimentation had raised the bottom of the riverbed. Declaring that no trees would be cut, the Premier called for dykes in downstream sections of the river to be strengthened, and put in place a plan to move more than 2.5 million residents out of the Jingjiang flood diversion area.

Let me pause here, to tell you of an old story about the Jingjiang flood diversion area, a section of the Yangtze between Zhicheng City in Hubei Province and Chenglingji City in Hunan Province. Originally, nature controlled floods. Along the Yangtze, in both the middle and downstream stretches of the river, several big natural lake systems assume the role of flood diversion, retention, and discharge. People living along Lake Dongting – one of those natural lake systems that acted as a flood basin of the Yangtze – had both a house and a boat. Typically, they would live in the house, but would move to their boats when the floods arrived and then return when the waters receded.

The boat people lived like that for thousands of years, until the Ming dynasty. But around 1600, Premier Zhang Juzheng from Hubei had a lofty idea to make more farmland for his citizens. He built dykes along the north side of the Yangtze, which pushed the flood water south into Dongting Lake. In the 1950s through to the 1970s – in Mao's era – people in Hunan (a name that means 'South Side of the Lake') started to build dams to create more farmland out of what had been the Dongting Lake. The Yangtze's diversion areas became smaller and smaller.

This was the situation Premier Zhu faced in 1998.

And so he issued a policy to forcibly resettle people from the Three Gorges Project region to areas far away – in stark contrast to previous policies.

Navigation

In the history of China, the Yangtze River is usually referred to as the Golden Waterway. Our former Premier, Zhou Enlai, used to say, 'if one damages this Golden Waterway, one could be called a criminal to the Nation.'

Nowhere in the world does anyone think a dam would be good for shipping. Do we see dams on the main waterways of the Rhine or the Mississippi?

In 1988, I interviewed an official who headed the Yangtze Shipping and Transportation Department. I asked him what he thought about the dam being used as a way to increase navigation. He forced a smile and said: 'Using only a small portion of the Three Gorges Project budget, the Transportation Department could meet the goals declared by the Project. This is the Party's decision. I have no choice but to obey.'

But, he told me, that when they signed the agreement to proceed with the dam, they took care to ensure the Project authority would provide a shiplift. What has happened since then? Has the project improved navigation on the Yangtze?

Since construction began, shipping in the middle current of the river has been cut off for 11 years. Several million passengers have passed through the dam and spent three hours in the five stages of water locks. Construction on the shiplift has only just begun, years after it was originally promised, and is now slated for completion in 2015.

In the dry season, docks downstream suffer, as ships are held up because of the low water level. Once, from November 2004 to June 2005, traffic problems lasted as long as 160 days – with shipping completely halted for 67 days.

These extended delays have led to the creation of a new word in Chinese: *fan ba*, which means 'dam turning over.' The popular term derives its meaning from the long wait times – up to dozens of hours – that affect cargo ships arriving at the dam during the Yangtze's busy seasons.

The solution is to bypass the dam! Goods are now transferred from the delayed ships to trucks for passage around the dam by road. Trucked goods are then loaded onto different ships to continue their journey.

Resettlement

How many people have been forced to relocate for Three Gorges Project?

Initially, Three Gorges Project supporters tried to lower the resettlement figure to make the dam more palatable. But once it was approved, the figure quickly rose.

Year	Resettlement figures
1992	725,500
1994	1,130,000
2004	1,300,000
2009	4,000,000*

Not only has the figure changed, but so too has the resettlement policy.

In the 1980s, the so-called trial resettlement programs were guided by two key goals: first, the relocation of people to higher ground in the same district, and second, migrants would be encouraged to find shelter with friends and family. Authorities then planned to allocate funds for migrant resettlement to the local governments of resettlement areas. These governments would

* The official figure in 2010 for the number of residents resettled by the dam is 1.4 million. But because many environmental problems have surfaced, the government has begun to resettle millions more. These subsequent resettlement programs have been done in the name of 'urbanization' or 'employment' programs, rather than because of the dam itself.

also be responsible for the building of new houses, and for providing farmland or factory jobs for migrants.

The problems with the resettlement plan, however, are first, the ecology in the Three Gorges is too fragile to cope with the resettlement of so many people; and second, it was incredibly risky to entrust local officials with the administration of resettlement funds without proper oversight mechanisms in place.

After the 1998 flood – and in light of the many problems with the resettlement operations, particularly with the deterioration of the environment along the Yangtze – Premier Zhu abandoned the policy of resettling all Three Gorges Project migrants in the same area in favour of a policy known as distant resettlement. Residents were soon moved to other parts of the country in an effort to relieve the pressure on the increasingly fragile environment in the reservoir area.

Since then, migrants have been relocated to 11 provinces and municipalities, including Shanghai, Guangdong, Fujian, Jiangsu, Zhejiang, Shandong, and Hubei.

Since the 1990s, migrants have not ceased their protests – both in public demonstrations and written petitions to higher authorities. Unfortunately, these protests have been met with force. No one knows how many of the migrant representatives have been sent back with a police escort, have been jailed or beaten so severely they are now afflicted with a disability.

Last year, Chongqing announced a huge new resettlement plan for its vast municipality – a plan that would result in the resettlement of some four million people, many of whom were originally moved to make way for the dam.

What did critics of the dam suggest when the Three Gorges Project was in its infancy? What unforeseen problems now plague the project, but which nobody will admit exists?

1. Water pollution

Neither the feasibility study in 1986 nor the one in 1990 mentioned water pollution, even though water pollution would affect not only the reservoir, but the Yangtze's tributaries as well. Neither of these pollution scenarios was included as costs in the project's budget.

As construction proceeded, an effort was made to clear polluted sites from the areas to be flooded. Factories, mines, hospitals, residential buildings, tombs – all of them became a 'second category' of pollution sources. Meanwhile, industrial waste and domestic sewage continued to be discharged into the reservoir without effective regulation. Although a number of pollution treatment factories were built along the reservoir, they rarely operate, as their operating costs are higher than the cost of construction.

The most polluted area is upstream, in the tributaries. According to the recent news report The Five-Year Plan for Water Pollution in the Three Gorges Reservoir Area and Upstream Areas states that the risk of water pollution by heavy polluting industries in the Three Gorges reservoir remains a problem. Due to the presence of industries that harvest raw materials, water pollution problems are common in various parts of the river – including phosphorus pollution caused by the mining of phosphate, heavy metal pollution caused by the mining of coal and arsenic contamination in specific areas. Some of the pollution has directly threatened the safety of drinking water.

2. Geological disasters

Right from the beginning of the project, critics highlighted the threat of geological disasters – but none of their estimates reflected the true gravity of the situation.

For instance, a crack was discovered in the concrete dam itself – as wide as an adult's hand stretching from top-to-bottom. Is it because of the quality of the cement or the pouring process?

In addition to repeated and variously sized mud-rock flows and landslides, a distortion and twist in the shiplock has developed. Engineers have discovered that the shiplock's gate frame doesn't match the gate. What kind of force would be so powerful to cause this distortion? Only the earth!

The distortion – which happened in the five-step shiplock – means the Earth's crustal plate is moving. Even the Three Gorges Project authority has admitted that this is a very serious problem. But no one knows why it happened and information about the issue is not available to the Chinese public.

3. Weather impact

Sleet, heavy storms and massive droughts, which were once rare along the Yangtze, its tributaries and south China, have all occurred, since the Three Gorges dam was initiated and in the latter half of the first decade of 2000.

And no one dares to admit that the pressure the earth must bear annually beneath 22 billion tons of water during the raising of the reservoir may induce an earthquake.

And now that sand is being blocked behind the dam, clear, sediment-free water has been washing out of the dykes downstream of the project.

In 2008, a rat infestation in the Dongting

Lake district, downstream from the dam, occurred when dam operators restricted water flow on the river.

The biggest headache for authorities appears to be the 30-meter drawdown belt that encircles the reservoir's 5,300-kilometre perimeter. This is the no-man's land that appears when the reservoir is lowered from 175m to 145m every year in preparation for the floods.

In early November 2010, Huang Qifan, the mayor of Chongqing suggested reducing the reservoir draw-down level from its current 30 metres (from the 175 metre Normal Pool Level to the 145 metre flood control level) to only 10 metres (so, from the NPL of 175 meters to a 165 – metre flood control level) during the flood season, thereby sacrificing much of the flood control storage capacity.

4. Budget

In order to get approval, the department in charge of the dam has used the 'announce a little early on, then replenish unceasingly' financing method. This is what we call in Chinese, 'small bait fishes big fish.'

The budget for the Three Gorges Project in the 1980s feasibility study was 36 billion RMB. By the time the Three Gorges Project was approved by the National People's Congress in 1992, it was 57 billion RMB. And then the figure continued to rise: 75 billion in 1993, 96 billion in 1994 and now it sits at 204 billion RMB. But the truth is that the real cost of the Three Gorges dam is 600 billion RMB – just as those who opposed the dam calculated in 1989. (It appears that the earlier investments estimated by the Chinese government did not include interest and inflation – known as the 'static investment.' The last estimates of 204 billion and 600 billion RMB, however, do include interest and inflation and are known as the 'dynamic investment.')

In fact, only the cost of dam construction, the power transmission network and resettlement programs are considered 'in' the budget. The cost of moving relics, of environmental protection and the treatment of polluted water in the reservoir are not included in the official budget.

Meanwhile, those responsible for the Three Gorges Dam use money for the project to play the stock market, invest in Hong Kong, and trade in Beijing's real estate market. No one has tried to put a stop to this and no one has investigated the losses yet. It's an unspoken truth that a great deal of money from the Three Gorges budget has been siphoned off for corrupt ends, and spent on sedans, villas, and bribes.

Supplementary funds paid by the central government:

Pollution treatment – 10 billion (RMB)
Historical preservation – 3 billion (RMB)
Resettlement – 5 billion (RMB)
Treatment after disasters – 1 billion (RMB)
Six extra underground turbines – 2.25
    billion RMB
Fancy entertainment and bribes*

---

* The exact amount is unknown but it is believed to be billions of RMB.

On the Completion of the Three Gorges Project

In the 1950s, 60s, and even the 1970s, Chinese citizens would have wholeheartedly followed whatever the Communist Party leaders said. Back then, we were willing to sacrifice our way of life to build a "new and strong China," home to reclaimed terraced fields (which actually meant deforestation) and dams (we have built more than 84,000 large and middle-sized dams since the 1950s, with two-thirds of them now deconstructed or considered dangerous). But now, with the advent of the 21st century, the entire world is beginning to understand both the positive and the negative impacts of the industrial revolution. The result is a sort of re-adjustment of the relationship between mankind and nature.

When the West (which used to be a pioneer in the damming of rivers) stopped building dams, why did China continue to build such a massive dam on a river which is over-exploited, over-populated and intensely damaged?

China's undemocratic political system and crony capitalist economic environment have, together, paved the way for the Three Gorges Project. The project is causing irretrievable losses and great harm to ordinary people and the environment.

With dramatic changes in China's economy, massive amounts of money are now in the hands of the central government. The Three Gorges dam won't be the last of this type built in China either; the Yangtze won't be the last of China's man-made disasters; and the ordinary people – who are being hurt by this project – will not be the last group to suffer in this way.

Encouraged by the Three Gorges Project, politicians and greedy businessmen are already extending their power to other rivers, to the coal and metal mines, and recently, the rare earth mines.

Records set by the Three Gorges Project rank it as the largest dam project built on Earth, with the longest reservoir, the highest shiplift and the largest capacity for power generation. But China can also claim to have the most polluted air, the most frequent mine catastrophes and deaths, the most expensive administrative costs and the largest gap between poor and rich with the highest percentage of an illiterate population.

Twenty-five years have passed since the debate over the Three Gorges Project first began. Though many Chinese citizens now earn an income of the middle class, they've become part of a silent majority. We don't have the right to express ourselves freely, nor the right to criticize the government openly, let alone the ability to monitor, to supervise or curtail the actions of officials.

I really don't know who or what can save the beautiful Three Gorges. I don't know – in exchange for the luxury of promoting a "rising" China – how much the environment and ordinary people will pay.

But I do know that, in the end, only we can save China when more people start accepting their role as citizens.

Your interest in and concern about the Three Gorges Project, the Yangtze, and China's environment warms my soul and encourages me.